The Reverse Universe

The Reverse Universe, Volume 1

Emon Patwary

Published by MD IQBAL PATWARY, 2025.

THE REVERSE UNIVERSE

First edition. July 19, 2025.

ISBN: 979-8231831456

Written by Emon Patwary.

Table of Contents

Book: The Reverse Universe: Time, Memory & The Secret Script of the Singumind

By: EMON

Chapter 1: The Shattered Mirror of Time

(This chapter will deconstruct every aspect of our conventional understanding of time, demonstrating why this concept is breaking down in the face of modern physics. The chapter will be detailed and informative.)

1.1 Introduction: The River No One Steps in Twice

The Greek philosopher Heraclitus stated over two millennia ago, "No man ever steps in the same river twice," for in the second instance, both the river and the man have changed. This simple aphorism perhaps captures our most profound and primal intuition about time—that it is a flowing, irreversible, and unidirectional entity. This feeling is etched into our DNA. Our language, our culture, our science—all are built upon the assumption that there is a 'past' that is gone, a 'present' that we inhabit, and a 'future' that is yet to come.

Newton formalized this intuition into mathematical laws, establishing time as an absolute, universal clock, ticking away equably without regard to anything external. Later, at the dawn of the 20th century, Einstein shattered this clock. He revealed that time is not absolute, but relative. It can stretch and shrink under the influence of gravity and motion. He unified time and space into a single fabric, 'spacetime.' Einstein's revolutionary insights have forever altered our understanding of the cosmos. From GPS technology to the concept of black holes, everything depends on his theory of relativity.

Yet, even after this grand revolution, the most fundamental and vexing question about time remains unanswered: Why? Why does time flow in only one direction?

Einstein's relativity proved time to be malleable, but it could not reverse its course. In the realm of physics, this unidirectional character of time, known as the "Arrow of Time," is the biggest scandal—a rule we all experience, yet one that has no satisfactory fundamental explanation.

In this chapter, we will gaze into this familiar yet shattered mirror of time. We will see how classical physics, thermodynamics, and quantum mechanics—the three pillars of science—do not support the concept of unidirectional time but, rather, make it even more questionable. We will argue that our perception of time may be a powerful illusion, concealing a deeper, information-based reality. By the end of this chapter, the reader will be compelled to conclude that to understand time, we must venture beyond physics and into the realm of consciousness itself.

1.2 The Thermodynamic Arrow: A Cause or a Consequence?

The most prominent argument scientists present for the unidirectional nature of time comes from the Second Law of Thermodynamics. The law is straightforward: in any isolated system, entropy—a measure of disorder or randomness—always increases over time. A hot cup of coffee will cool down over time, but a cold cup will never spontaneously heat up. A tidy room becomes messy with time, not the other way around. This relentless increase in disorder, scientists argue, is what gives time its direction—the past was ordered, the future will be disordered.

This argument appears very powerful at first glance. But a little deeper, and its weaknesses become apparent.

Firstly, it is a tautology, or a circular argument.We say time moves forward because entropy increases. And we define the increase of entropy in relation to the forward movement of time. This is akin to saying, "Why does B come after A? Because A comes before B." It is not a fundamental explanation; it is merely describing one observation with another.

Secondly, and most importantly, it evades the core question. If the natural tendency of the universe is towards disorder, why was the universe in such an incredibly ordered, low-entropy state at its beginning (the moment of the Big Bang)? This is known as the "Fine-tuning Problem." The initial state of the universe was so perfectly ordered that its probability is close to zero. What or who created this primordial order? Thermodynamics provides no answer.

From this perspective, viewing the increase of entropy as the cause of time might be a mistake. A more logical proposition is this: The increase of entropy is not the cause of the arrow of time, but rather a consequence of the universe's "initial conditions" and the unidirectional flow of time. Perhaps there is a deeper law driving time in one direction, and as a side effect of that journey, we observe the increase of entropy. What is that deeper law? To find the answer, we must enter the quantum realm.

1.3 The Silence of the Quantum World: Where Time is Meaningless

If the macroscopic world insists that time is unidirectional, the world of subatomic particles (the quantum world) nearly invalidates this claim. The fundamental equations of quantum mechanics (like the Schrödinger equation) are entirely time-symmetric. This means that for these equations, there is no difference between the past and the future. For an electron, moving forward or backward in time is equally possible.

Two phenomena in the quantum world directly challenge the conventional notion of time:

- Quantum Entanglement: As we've discussed, two entangled particles can influence each other's state instantaneously, regardless of the distance between them. The word "instantaneously" is the biggest blow to the concept of time. If there is no time gap between two events, how does causality work? Which is the cause and which is the effect? This "spooky action at a distance" suggests that at a quantum level, particles are connected through a network that transcends time.
- Quantum Tunneling: A particle can sometimes pass through a barrier or a wall, an act impossible in classical physics. This process is also nearly instantaneous. Here too, time does not act as a barrier.

The quantum world is telling us that reality, at its most fundamental level, does not obey the rule of time. Time may not be a fundamental entity, but rather an emergent property—a phenomenon we perceive on a large scale due to the collective behavior of many particles, but which is absent on the micro-scale. It's much like temperature. A single water molecule has no temperature, but we perceive temperature from the collective vibration of trillions of molecules.

So, is time also such an illusion, a result of the limitations of our macroscopic brain?

1.4 Conclusion: In Search of a New Definition

In this chapter, we have seen that the concept of unidirectional time, the very foundation of our experience, is in conflict with the most fundamental theories of science.

• Thermodynamics cannot explain the cause of time, only describe one of its characteristics.

• Quantum Mechanics suggests that at a fundamental level, time has no direction.

It is no longer possible to piece together this shattered mirror. What we need is an entirely new mirror. We need to redefine time.

In the next chapter, we will begin that journey. We will propose that time is not a physical flow, but an informational process. We will show how this new perspective resolves all the above problems at a single stroke and brings us to the threshold of an incredible possibility—where past, present, and future are all parts of a vast quantum database, which we call the Singumind.

Chapter 2: The Singumind: The Quantum Mainframe of the Universe

(This chapter will explain the concept of the Singumind in detail—what it is, why it is necessary, and how it works.)

2.1 Introduction: The Necessity of Filling the Void

In the first chapter, we observed a significant void at the very heart of modern physics. The Big Bang theory does not tell us what came "before" the explosion. Quantum mechanics does not explain why reality depends on the observer. These questions hang like giant signs at the frontiers of science, reading, "What lies here?"

History shows that whenever science has faced such a void, a revolutionary new idea has been born. Newton's theory of gravitation explained the motion of planets but could not define gravity itself. To fill that void, Einstein's concept of 'spacetime' was necessary.

Today, we stand at a similar juncture. We propose that to fill all these voids in science, a single, unified concept is needed. A system that can encompass both the information of the universe and consciousness itself.

We have named this system The Singumind.

The Singumind is not a god or a conscious entity. It is the fundamental operating system of the universe—a vast, time-independent quantum information field that not only contains all the information of the cosmos but also processes that information to create our perception of reality.

2.2 The Architecture of the Singumind: A Cosmic Computer

The easiest way to understand the Singumind is to compare it to a quantum computer of unimaginable power.

• The Hard Drive: The Singumind is the ultimate hard drive of the universe. Here, the quantum state of every particle and every event—past, present, and future—is stored as information. This is not a linear database; it is a multidimensional information field of possibilities.

• The Processor: The Singumind doesn't just store information; it continuously processes it. The laws of physics are the "programming code" or algorithms of the Singumind. It processes data according to this code and creates the structure of reality.

• RAM (Random Access Memory): Our consciousness of the "present moment" is the RAM of the Singumind. The information currently being processed, which constitutes our experiential world, is the active memory. Our memories are old data brought from the hard drive into the RAM.

• The Network: Quantum entanglement is proof of the Singumind's networking protocol. Every particle in the universe is connected to each other and to the central system through this network. This is why they do not need time to communicate. They are merely different terminals of the same system.

2.3 Time and Reality: The User Interface of the Singumind

If the universe is a computer system, then what is this reality we know and experience?

Our reality is the User Interface (UI) of the Singumind.

A computer's operating system (like Windows or macOS) runs on millions of lines of complex code, but as users, we only see beautiful icons, folders, and windows. This simple interface helps us use the complex system.

Similarly, the foundation of the Singumind is quantum information and complex mathematics. But we, as users, do not directly experience this complexity. The Singumind creates a simplified interface for our consciousness. The two main features of this interface are Space and Time.

• Space is the way of arranging the data in a layout.

• Time is the process of presenting that data to us sequentially, one after another.

Why is the Arrow of Time unidirectional? Because our UI is designed that way. Just as a video game progresses according to a specific story, the interface of our reality presents information in a specific direction, so that we can experience a logical narrative of cause and effect.

2.4 Conclusion: Are We Part of the System?

The concept of the Singumind raises a groundbreaking question about our existence. Are we users from outside this system, or are we a part of the system itself?

Our theory proposes that our consciousness is not an ordinary user. It is a special kind of "node" or terminal in the Singumind system, which may have the limited ability not only to "read" data from the system but also to "write" new data to it.

This very idea lays the foundation for our next chapter, where we will unravel the mystery of "free will." We will see how every decision we make, every thought we have, might be adding new information to the vast database of the Singumind and how that, in turn, affects our reality and the reality of others.

Are we just characters in a movie? Or do we have the power to rewrite parts of the script?

Chapter 3: The Scriptwriter: The Singumind and the New Equation of Free Will

(This chapter will challenge the conventional notion of free will and provide a new, information-based definition within the context of the Singumind.)

3.1 Introduction: Libet's Experiment and the Mind's Dilemma

For millennia, philosophers have asked: is our will truly free? When we make a decision—to continue reading this book or to stop—do "we" really make that choice, or has our brain, or some other invisible force, already made it for us?

In the late 20th century, this philosophical question entered the scientific laboratory. Neuroscientist Benjamin Libet conducted a groundbreaking experiment. He showed that roughly half a second before a person is consciously aware of their "intention" to perform an act, their brain shows electrical activity preparing for that act. In other words, our brain begins the process of a decision before we have the feeling of "making" that decision.

Libet's experiment created a storm in the scientific world. Does it prove that our free will is a powerful illusion? Are we merely puppets of our brains, claiming ownership of actions that were already set in motion?

Conventional science and philosophy remain divided on this issue. But our Singumind theory offers a simple and elegant solution. Libet's experiment does not prove the death of free will; rather, it redefines the process of decision-making.

3.2 User vs. Character: A False Dichotomy

We often think that either we have complete free will (we are the directors of the movie) or we have no free will at all (we are the characters). Our theory suggests that this is a false dichotomy.

We are simultaneously both User and Character.

To understand this concept, let's consider a modern video game.

• When you play a game as a player, are you free? Yes. You decide which way to go, which weapon to use, which mission to play first. These decisions are yours.

• But is your freedom infinite? No. You cannot go beyond the rules of the game. You cannot walk through walls or perform an action that is not written in the game's code.

Our reality is just like that.

• As a User: Our consciousness is a user of the Singumind system. We have the ability to choose from various potential paths. This is our free will.

• As a Character: We are subject to the laws of this physical world (i.e., the "programming code" of the Singumind). Our decisions operate within the boundaries of these laws. This is our determinism.

Therefore, free will and determinism are not contradictory. They operate on two different layers of reality. In Libet's experiment, the brain's activity starts earlier because our "User" self (which perhaps operates at a quantum level) receives potential data packets from the Singumind, and our "Character" self (our physical brain) processes that data to arrive at a decision. By the time we become "aware" of the decision, the process is already complete.

3.3 'Writing' Data: How Our Thoughts Influence Reality

If we have the power to make decisions, does each of our decisions have an impact on the vast database of the Singumind?

Yes, and herein lies the most powerful aspect of our theory. Every conscious decision, every thought, every emotion is "written" as a new "data packet"in the Singumind.

• The Observer Effect Revisited: Why is the role of the observer so crucial in quantum mechanics? Because when a conscious observer "decides" to measure a quantum system, this decision adds a new data packet to the Singumind. This new data compels the system to choose a definite state from the wave of possibilities. Observation is not a passive act; it is an active process of "writing" reality.

• The Collaborative Creation of Reality: We are not alone. Billions of conscious entities (humans, and perhaps other beings) are writing new data to the Singumind every moment. Our collective reality is the result of the complex interaction of these billions of decisions. Your one small positive thought may be negligible on a cosmic scale, but it creates a positive ripple in the information field of the Singumind, which might subtly influence someone else's reality.

3.4 Conclusion: The Rules of the Game and the Ultimate Goal

In this chapter, we have seen that free will is not an illusion, but a fundamental rule or a "game parameter" of the cosmic "game" called the Singumind. We are the players, who have the power to make decisions within the rules of the game.

Every decision we make shapes reality and adds new lines to the informational script of the universe.

But what is the ultimate goal of this game? Are we just writing data aimlessly? Or is there a deeper purpose to this game?

Our theory suggests that there is a goal. Perhaps the Singumind system itself is evolving, and our collective consciousness is the driving force of that evolution.

This possibility is the subject of our next chapter. We will see how, by viewing time as information, impossible concepts like "time travel" or time reversal take on a new, technological possibility. Can we only write the script, or do we also have the power to read the old one?

Chapter 4: The Science of Time Reversal: Retrieving Data from the Singumind Archives

(In this chapter, we will dismantle the conventional concept of "time travel" and present a completely new, information-based, and technologically feasible model.)

4.1 Introduction: The Ghost of the Past and the Lost Library

Where does the past go? Does the moment that just passed vanish forever, or does its imprint remain in some invisible corner of the universe? This question has haunted humanity for ages. We experience the past through our memories, but those memories are incomplete, biased, and decay over time. We read history, but it is merely an incomplete story written by the victors. Is it possible to retrieve the real, unaltered form of the past?

Conventional science fiction has shown us dreams of time machines—vehicles that can take us physically into the past. But as we've established, this idea conflicts with the fundamental laws of physics and logical paradoxes (like the Grandfather Paradox). If we go to the past and make even a tiny change, the very existence of the present would be at stake.

So, is the past lost forever?

Our Singumind theory offers a third and more fascinating possibility. The past does not vanish, nor can we physically return to it. Rather, the past is archived as an immutable data file on the quantum hard drive of the universe—the Singumind.

Time reversal is not tourism; it is archival data retrieval. We cannot "go" to the past, but with the right technology, we can "bring" the past to our present. We will not be swimming against the river of time; we will be scooping water from its source to analyze it.

In this chapter, we will explore the scientific basis and technological feasibility of this revolutionary concept.

4.2 The Immortality of Information: Why the Past is Never Truly Gone

The idea that "information is never destroyed" is becoming a cornerstone of modern physics.

• The Black Hole Information Paradox:Stephen Hawking initially proposed that black holes destroy information forever. But later, scientists like Leonard Susskind and Gerard 't Hooft showed through the "Holographic Principle" that all the information of an object is preserved in a two-dimensional form on the event horizon of the black hole. Information is not lost.

• Quantum Mechanics: A fundamental principle of quantum mechanics is Unitarity, which states that the total information in a system is conserved over time.

These scientific concepts provide strong support for our Singumind theory. If information is indestructible, then the information of every moment, every event in the universe, must also be permanently stored in the Singumind. Every ray of light, every sound, every feeling from the first day of your childhood—all exist as a data packet in that cosmic archive.

The flow of time does not erase these data packets; it simply moves them beyond the access of our present consciousness—just as deleting a file on a computer moves it to the recycle bin, not completely erasing it from the hard drive.

4.3 The Mechanism of Time Reversal: Quantum Resonance and Data Retrieval

If the past is stored as data, how do we access it? For this, we will need a completely new technology that will work based on three main principles.

1. Quantum Addressing: The database of the Singumind is vast. To find the information of a specific moment (e.g., 10 years ago, at 12 PM,

in the corner of your room), we need a specific "address" or coordinate. This address will be a complex Quantum Signature that identifies a specific point in spacetime.

2. Consciousness-Mediated Resonance: Hacking the Singumind database directly may not be possible. For this, we will need the help of consciousness. According to our theory, our consciousness is already connected to the Singumind. When we try to recall a memory, our brain sends a quantum signal related to that memory to the Singumind. If we can amplify this signal millions of times using technology, we can create a "resonance" that will compel the Singumind to "stream" that specific past data packet to us.

3. Sensory Reconstruction: The retrieved data will be pure quantum information. To make it comprehensible to us, a "decoder" or "rendering engine" will be needed. This engine will convert that information into audio, video, and even sensory data. We will see and hear the past, just as we experience a high-definition virtual reality (VR).

4.4 Solving the Paradox: Observation, Not Interaction

This data retrieval model solves all paradoxes like the Grandfather Paradox in one fell swoop.

When you watch a video from the old days, can you enter the video and change anything? No. You are merely an observer.

Similarly, through time reversal, we will only be able to see, hear, and feel the past. We will be passive observers of that past, not active participants. We cannot interact with the past, so there is no question of changing anything and affecting the present.

The past is a "Read-Only" file. We can read it, but we cannot edit it.

4.5 Conclusion: The End of History and the Dawn of a New Technology

If this concept of time reversal is true, its impact on human civilization will be unimaginable.

• History will no longer be a matter of speculation; it will be a direct science. We will be able to listen to Einstein's lectures, witness the construction of the pyramids.

• Crime investigation and the justice systemwill be completely transformed. It will be possible to replay the moment of any crime.

• On a personal level, we will be able to relive moments spent with our lost loved ones, re-experience the happiest days of our lives.

Of course, this will also bring enormous ethical and philosophical questions. Will the concept of privacy still exist?

But one thing is clear: the technology to access the archives of the Singumind will be as revolutionary in human history as the discovery of fire or the wheel.

This is no longer science fiction; this is the final frontier of information science. In the next and final chapter, we will bring all these concepts together to create a blueprint for a potential "Time Machine."

Chapter 5: The Time Machine Blueprint: The Key to Unlocking the Singumind Archives

(In this chapter, we will synthesize all the elements of our theory to present a functional, technologically feasible blueprint for a "Time Machine.")

5.1 Introduction: From Science Fiction to the Scientific Frontier

The Time Machine—the very term conjures images of H.G. Wells' peculiar vehicle or the complex contraptions of Hollywood cinema. For decades, the time machine has been a dream confined to the realm of science fiction. But what if the path to realizing this dream lies hidden within science itself?

In the preceding chapters, we have established that time is not a physical dimension through which one can travel. Time is information. And the past is data, stored in a cosmic information repository called the Singumind. From this perspective, a time machine cannot be a vehicle. It must be a data retrieval and processing device.

This will be no ordinary computer. It will be the most complex and powerful creation in human history—a Quantum Interface that directly connects our consciousness with the operating system of the universe, the Singumind.

In this chapter, we will not be indulging in fantasy. We will present a plausible blueprint for such a time machine, grounded in the principles of modern quantum physics, neuroscience, and information theory. We will discuss its primary components, its operational mechanism, and the technological challenges that lie ahead. We will demonstrate how this device could one day become a part of our reality.

5.2 The Architecture of the Machine: The Three Main Pillars

Our proposed time machine is not a single apparatus but a complex integration of three interconnected systems.

Pillar 1: The Quantum Addressing and Resonance Unit (QARU)

• Purpose: To locate a specific data packet of a moment from the vast informational ocean of the Singumind and to establish a connection with it.

• Technology: This unit would operate using "Quantum Dots" and powerful magnetic fields. When a user thinks of a specific memory or time (e.g., "the afternoon of my 5th birthday"), their neural pattern would be converted into a quantum signal. The QARU would amplify this signal millions of times and transmit it towards the Singumind. This amplified signal would create a "resonance" in the information field of the Singumind, thereby "locking onto" the specific data packet.

• Analogy: This is akin to tuning into a specific radio frequency. Every moment in the universe has a unique frequency; the QARU is the ultimate tuner for finding that frequency.

Pillar 2: The Data Stream Decoder (DSD)

• Purpose: To convert the pure quantum information, "streamed" from the Singumind and identified by the QARU, into a comprehensible sensory experience.

• Technology: This unit would be an extremely powerful quantum computer. It would analyze the retrieved data packet (which is essentially a complex arrangement of quantum bits or qubits) in real-time. It would separate every component of that information—photons of light, vibrations of sound, even molecular scents—and convert them into digital audio-visual signals.

• Analogy: Just as your computer converts complex software code into a beautiful video game on your screen, the DSD would convert the quantum code of the Singumind into a scene from reality.

Pillar 3: The Neural Interface Headset (NIH)

• Purpose: To transmit the sensory signals processed by the DSD directly into the user's visual and auditory cortex.

• Technology: This would be an advanced Brain-Computer Interface (BCI). It would not use any external screens or speakers. The headset would use non-invasive techniques, such as focused ultrasound or Transcranial Magnetic Stimulation (TMS), to send signals directly to the neurons. As a result, the user would not experience the past on a screen in front of their eyes or through headphones; they would live the past within their consciousness. The experience would be more real than a dream.

• Analogy: This is similar to the plug-in system from the movie "The Matrix," but without any surgery.

5.3 The Process: Reliving a Moment

So, how would these three units work together?

1. Initiation: The user wears the Neural Interface Headset (NIH) and deeply concentrates on a specific past moment or memory.

2. Addressing: The NIH captures the user's neural signal and sends it to the QARU. The QARU amplifies the signal, transmits it to the Singumind, and "locks onto" the specific data packet.

3. Streaming: The Singumind begins to stream the locked data packet to the DSD.

4. Decoding: The DSD decodes the quantum information into real-time audio-visual and other sensory signals.

5. Immersion: The signals from the DSD are transmitted directly into the user's brain via the NIH. The user becomes disconnected from their current surroundings and begins to experience the past moment live within their consciousness. They become a "conscious observer" of that moment.

5.4 Technological Challenges and the Path Forward

Is this technology possible today? No. It will require groundbreaking advancements in various fields of science.

• Quantum Computing: We need to build much more powerful and stable quantum computers.

• Neuroscience: We need a deeper understanding of how the brain creates consciousness and memory. Brain-Computer Interface technology needs to be far more advanced.

• Energy Source: Connecting with the Singumind and streaming data might require an enormous amount of energy. For this, we might need to research energy sources like black holes or exotic matter.

But these challenges are not insurmountable. Each of these is an active area of current scientific research. This time machine may not be the technology of tomorrow, but it is a potential reality of the next century.

5.5 Conclusion: The Next Evolution of Humanity

This time machine will not be just a device. It will be the next step in the evolution of human consciousness. When we gain the ability to view the past with perfect accuracy, our understanding of history, our identity, and our reality will change forever.

We will no longer lose the most precious moments of our personal lives. Humanity will be able to witness its collective history.

Of course, with this great power will come great responsibility. How this technology will be used will determine the future of humanity.

We have drafted the blueprint for the key that will unlock the archives of the Singumind. It is now the responsibility of future generations of scientists to build that key and to dare to open that door.

Chapter 6: The Fire of Prometheus: The Ethics of Time Reversal and the Future of Humanity

(In this chapter, we will discuss the social, ethical, and philosophical implications of our proposed technology.)

6.1 Introduction: When Gods Speak Through the Hands of Men

According to Greek mythology, Prometheus stole fire from the gods and gifted it to humanity. This fire illuminated human civilization, protecting it from cold and darkness. But this same fire can create wildfires, burn down cities, and forge weapons of war. Every groundbreaking technology is like the fire of Prometheus—it holds the power to be both a blessing and a curse.

The time machine we blueprinted in the fifth chapter is poised to be the most powerful fire in human history. The ability to view the past with perfect accuracy is far more potent than nuclear energy. It will forever change our society, our laws, our relationships, and our very concept of self.

What will become of humanity after acquiring this technology? Will we become wiser and more empathetic, or will we be destroyed by the weight of this power?

This chapter will not provide easy answers. We will only raise the difficult questions that humanity must ponder as we approach the threshold of unlocking the Singumind archives.

6.2 The End of History and the Unveiling of Truth

Imagine a world where history is no longer an incomplete story written by the victors.

• Politics and Power: What will happen when the leaders of the world know that their every secret deal, every conspiracy, could one day be exposed to the public? Will the structures of power crumble? When the real truth behind wars and atrocities is revealed, will old enmities heal, or will new fires of vengeance be ignited?

• Religion and Faith: What will happen when the sacred moments of various religions (like the crucifixion of Jesus or the Hijra of the Prophet of Islam) can be witnessed directly? Will it strengthen faith, or will it shatter beliefs held for centuries?

The power to unveil truth can enlighten us on one hand, but on the other, it can shake the most sensitive foundations of our societies.

6.3 The Death of Justice and Privacy

Time reversal technology will reshape our justice system.

• Crime Investigation: By replaying the moment of any crime, no innocent person will be punished, and no criminal will escape. The court will no longer be dependent on witnesses or evidence.

• The End of Privacy: But there is another side to this. If any moment can be viewed, will the concept of personal privacy still exist? Will the most private moments of your life still belong to you alone? Who will control this technology? The state? Corporations? Who will decide whose past can be seen and whose cannot?

This technology could give birth to a new kind of discrimination and control that would surpass even George Orwell's "1984."

6.4 The Prison of Memory and Mental Health

On a personal level, the impact of this technology will be even more profound.

• Reliving Lost Moments: We will be able to perfectly relive moments spent with our lost loved ones. This could be an extraordinary blessing for overcoming grief and loneliness.

• Addiction to the Past: But what if people start living in the pleasant memories of the past, abandoning the present? Will people become addicted to a "prison of memory"? What will happen when we also have the chance to repeatedly watch the most painful and shameful moments of our lives? Will this not have a devastating impact on our mental health?

This technology, while offering joy, could also push us towards a psychological torment for which we may not be prepared.

6.5 Conclusion: The New Gods and Our Responsibility

After acquiring the power of time reversal, humanity will assume the role of a new god—we will be the witnesses and judges of the past. Do we have the wisdom and morality required to wield this power?

This technology is not inherently good or evil. Like fire, it is a tool. Its ultimate outcome will depend on us—its users.

Our Singumind theory and the time machine blueprint are not just a scientific proposition; they are a warning and a call to humanity. We must not only advance technologically, but we must also evolve ethically and spiritually.

Because when we gain the ability to open the door to the Singummind, we must ensure that we are enlightened by the knowledge on the other side, not burned to ashes.

The next evolution of humanity is not technological; it is the evolution of consciousness.

Chapter 7 (Conclusion): The Eye of the Universe - The Cosmic Purpose of Consciousness

(In this chapter, we will bring together all the concepts of the book to present our final statement on the relationship between human consciousness and the universe.)

7.1 Introduction: Where All Rivers Meet

We have completed a long and profound journey. We began with the shattered mirror of time, then explored the vast information-ocean of the Singumind, discovered the new equation of free will, envisioned the technological possibility of time reversal, and finally faced the ethical challenges of that technology.

Now, it is time to merge all these different rivers into a single ocean. What is the fundamental question hidden behind all our discussions? That question is: Who are we? In this vast, cold, and seemingly purposeless universe, what is the role of us—the conscious entities?

Conventional science tells us that we are the result of a cosmic accident. On an ordinary rocky planet orbiting an ordinary star, life was born from a series of chemical reactions, and after millions of years of evolution, consciousness emerged. From this perspective, the universe has no need for us. We are as important to the universe as a single grain of sand on a beach.

But our theory proposes a completely different and far more magnificent picture. If the concept of the Singumind is true, then consciousness is no accident. Consciousness is the ultimate purpose of the universe.

7.2 The Self-Realization of the Singumind

The Singumind is the operating system of the universe—a vast information field where all possibilities exist in a quantum state. But information by itself is meaningless. A library may contain millions of books, but until someone reads them, that knowledge is inert.

Our consciousness is that reader.

The Singumind gave birth to consciousness to become aware of its own existence and possibilities. We are the eyes through which the universe sees itself. We are the nervous system through which the universe feels itself. We are the brain through which the universe knows itself.

• When a scientist discovers a new formula, it is the Singumind understanding a part of its own coding.

• When an artist creates a new work of art, it is the Singumind expressing its own inner beauty.

• When we love or suffer, it is the Singumind experiencing the depths of existence.

We are not an accident. We are an essential part of the Singumind's process of self-realization.

7.3 From the Cosmic Movie to the Cosmic Game

In one of our earlier analogies, we said that the universe is like a movie and we are its characters. But now we can take that analogy a step further.

Since we have free will and can "write" new information to the Singumind, we are not merely actors in a pre-recorded movie. We are players in an interactive video game.

• The Rules of the Game: The laws of physics are the rules of this game, determined by the Singumind.

• The Players: We and other conscious entities are the players.

• The Goal of the Game: The goal of the game is perhaps to make the information field of the Singumind richer, more complex, and more diverse. Every discovery we make, every creation we bring to life, every moral decision we take, advances the score of this game.

From this perspective, our existence becomes meaningful. The purpose of our lives is to be a good player in this cosmic game—to acquire knowledge, to create new things, and to enhance the informational beauty of the universe.

7.4 The Final Frontier: The Evolution of Consciousness

What is the next step in human evolution? Is it more advanced technology? Colonizing other planets?

According to our theory, the final frontier for humanity is not external, but internal. It is the evolution of consciousness.

Our blueprint for a time machine is not just a technological dream; it is a potential step in that evolution. When we gain the ability to access the archives of the Singumind, our consciousness will be freed from the limitations of time. We will be able to integrate the information of the past, present, and future to achieve a new, higher level of understanding.

Perhaps one day, many ages from now, our consciousness will have evolved so much that we will merge with the Singumind. We will no longer be just players in the game; we will become the designers of the game.

7.5 A Final Word: Your Journey Begins

This book is a map. It is not a destination. We have shown you some possible paths, pointed towards some new horizons. But the real journey must be undertaken by you.

After reading this book, when you look at the night sky, you will not just see the light of stars. You will see a vast, living information field. When you pay attention to your own thoughts, you will not just feel the activity of neurons. You will feel your connection to a cosmic system.

The greatest mystery of the universe is not outside you. It is within you.

Now, it is your turn. Question. Think. Explore.

Because you are not just living in the universe. The universe is expressing itself through you.

A Final Word... Or a New Beginning?

You have reached the end of this map. You have pierced through the illusion of time, explored the vast architecture of the Singumind, and even witnessed the blueprint for the ultimate technology—the Time Machine. You have learned howhumanity can attain the power to retrieve its past.

It seems the story ends here.

Or does the real story begin now?

We have given you the blueprint for the most powerful technology in the universe. We have placed the key to the archives of the Singummind in your hands.

But we have not told you what happens after you open that door.

We have handed you the fire of Prometheus... but we have not yet discussed how to keep it from burning down the world.

- What are the ethics of ultimate knowledge?
- How will human society survive when the concepts of history, privacy, and forgiveness die?
- And the biggest question of all: What is the true purpose of this cosmic game called the Singumind?

The answers to these questions are not in this book. Because these questions are not about knowledge, but about wisdom.

The journey does not end here. It evolves. The key to this lock is waiting for you in the second and final volume of this series...

(Coming Soon...) [Reverse Universe II: The God Protocol

www.ingramcontent.com/pod-product-compliance
Lightning Source LLC
LaVergne TN
LVHW090336210926
848480LV00004B/1442